THE NUCLEAR FUSION

Unlocking the Power of the Stars A Guide to Understanding Nuclear Fusion

By
Peter Whiteside

Table of contents

INTRODUCTION

FISSION AND FUSION

Fission and Fusion are two separate processes that are both crucial to the generation of energy. Fission is the process of splitting an atom into two or smaller atoms, releasing a huge quantity of energy in the form of radiation. Fusion is the process of fusing two tiny atoms into one bigger atom, thereby releasing a substantial quantity of energy in the form of radiation.

Fission is the method utilized in nuclear power reactors to create energy. It is commonly referred to as "atom-splitting". In a nuclear fission process, a big atom such as uranium is divided into two or smaller atoms, termed fission products. As the atoms break, they release a huge quantity of energy in the form of radiation. The radiation is then utilized to heat water, which generates steam. The steam is used to spin turbines that create power.

The process of fission is powered by neutrons, which are particles with no electrical charge. Neutrons may penetrate the nucleus of an atom and cause it to divide. The splitting of the atom releases energy in the form of gamma rays, neutrons, and other particles. This process is known as nuclear fission.

Fusion is the process utilized in stars to create energy. In a nuclear fusion process, two little atoms such as hydrogen are joined to generate a bigger atom, such as helium. As the atoms combine, they release a huge quantity of energy in the form of radiation. This radiation is subsequently utilized to create energy in a power plant.

The process of fusion is fueled by protons, which are particles with a positive electrical charge. When two protons hit, they may generate a new nucleus. This process is known as nuclear fusion.

Nuclear fission and fusion both provide significant quantities of energy, but they have distinct downsides. Fission is a somewhat slow process, and it creates radioactive waste that must be carefully kept for thousands of years. Fusion is a considerably quicker process, but it creates relatively little radioactive waste.

Fission and fusion are both fundamental to the creation of energy. They are both used to create energy in power plants, and they both have their benefits and limitations. As technology progresses, scientists are attempting to make both procedures more efficient and safer.

With these four instances, you will grasp the difference between fusion and fission

Example 1: Fission is the process of breaking an atom, releasing energy in the form of radiation. Fusion is the process of fusing two or more atoms, also releasing energy in the form of radiation.

Example 2:Fission is used in nuclear power plants to create electricity, whereas fusion is utilized in stars to produce energy.

Example 3:The process of fission is driven by neutrons, whereas the process of fusion is driven by protons.

Example 4:Fission creates radioactive waste, but fusion produces relatively little radioactive waste.

Chapter 1

FUSION BREAKTHROUGH

More energy out than in. For 7 decades, fusion scientists have followed this elusive aim, known as energy increase. At 1 a.m. on 5 December, researchers at the National Ignition Facility (NIF) in California finally accomplished it, concentrating 2.05 megajoules of laser light onto a small capsule of fusion fuel and causing an explosion that created 3.15 MJ of energy—the equivalent of approximately three sticks of dynamite.

"This is incredibly exciting, it's a significant breakthrough," says Anne White, a plasma physicist at the Massachusetts Institute of Technology, who was not involved in the study.

Mark Herrmann, who heads NIF as the program director for weapons physics and design at Lawrence

Livermore National Laboratory, says it feels "wonderful," adding: "I'm proud of the team."

The discovery, released today by officials at the U.S. Department of Energy (DOE), provides a shot in the arm for fusion researchers, who have long been chastised for overpromising and underdelivering. Fusion presents the tempting promise of copious, carbon-free energy, without many of the radioactive hassles of fission-driven nuclear power. But getting hydrogen ions to fuse into helium and produce energy demands temperatures of millions of degrees Celsius—conditions that are challenging to attain and maintain. The NIF result demonstrates it is achievable, at least for a fraction of a second. "Three MJ is a heck of a lot of energy. It suggests something is working," says plasma physicist Steven Rose of Imperial College London.

Despite the hype, fusion power plants are still a distant dream. NIF was never meant to generate electricity commercially. Its principal mission is to make tiny thermonuclear explosions and give data to assure the U.S. stockpile of nuclear weapons is safe and dependable. Many academics think furnace-like tokamaks are a superior design for commercial electricity because they can maintain longer fusion "burns." In a tokamak, microwaves and particle beams heat the fuel while magnetic fields trap it. "The difficulty is to make it sturdy and simple," White explains.

However, the leading tokamak device, the ITER reactor under development in France, is anything from simple. It is substantially over budget, long delayed, and will not approach breakeven until the late 2030s at the earliest. With NIF's current accomplishment, proponents of such laser-based "inertial fusion energy" will be lobbying for funding to investigate if they can compete with the tokamaks.

The $3.5 billion NIF launched its "ignition" campaign in 2010. Its laser, situated in a facility the size of three U.S. football fields, generates a strong, nanoseconds-long infrared pulse divided into 192 beams that are converted to ultraviolet light. The beams are focused on the target—a gold can the size of a pencil eraser carrying a peppercorn-size fuel capsule. Heated to millions of degrees, the gold generates x-rays that melt the diamond shell of the capsule. The blasted diamond implodes the fuel, compressing and heating it.

If the compression of the fuel is symmetrical enough, fusion reactions begin in a center hot point and move smoothly outward, with the heat from fusion igniting additional burning. That self-sustaining burn is what constitutes ignition, and after more than a decade of labor NIF scientists claimed they had accomplished that milestone after a blast in August 2021 generated 70% of the input laser energy. But NIF's underwriter, DOE's National Nuclear Security Administration, defined NIF's target as an energy gain larger than one—the level it reached last week.

Going that additional mile wasn't easy. After the August 2021 shot, the NIF team realized it couldn't duplicate it. Using a smooth diamond capsule proved to be key: The one from August 2021 had been the most flawlessly smooth and spherical they'd manufactured. "We had to learn how to manufacture the capsules better," Herrmann adds. They also made the capsule somewhat thicker, which supplied more velocity for the implosion but needed a longer, more intense laser pulse. So they modified the laser to squeeze out extra power, boosting the energy from 1.9 MJ to 2.05 MJ.

A blast in September generated 1.2 MJ, demonstrating to the NIF researchers they were on the right road, but the symmetry was poor: The fuel was crushed into a pancake rather than a tight ball. By modulating the energy among the laser's 192 beams, they were able to produce a more circular implosion, and last week they finally struck the jackpot. "The physics phenomena has been demonstrated," says Riccardo Betti of the Laboratory for Laser Energetics at the University of Rochester.

If gain meant generating more output energy than input electricity, however, NIF fell well short. Its lasers are inefficient, needing hundreds of megajoules of power to create the 2 MJ of laser light and 3 MJ of fusion energy. Moreover, a power plant based on NIF would need to elevate the repetition rate from one shot per day to around 10 per second. One million capsules a day would need to be created, filled, positioned, blasted, and swept away—a massive technical problem.

The NIF plan has additional inefficiency, Betti notes. It depends on "indirect drive," in which the laser shoots the gold container to create the x-rays that initiate fusion. Only around 1% of the laser energy gets into the fuel, he adds. He supports "direct drive," a strategy studied by his group, where laser beams shoot directly into a fuel capsule and deposit 5% of their energy. But DOE has never supported a program to develop inertial fusion for power production. In 2020, the agency's Fusion Energy Sciences Advisory Committee advised it should, in a study co-authored by Betti and White. "We need a new paradigm," Betti adds, but "there is no obvious way how to accomplish it."

Now that NIF has cracked the nut, researchers anticipate laser fusion may acquire legitimacy and additional money may follow. After the hard struggle to get here, Betti joked about transferring the baton. "This is a crucial first step," he adds. "We've done it now, so I can retire."

Chapter 2

IS NUCLEAR FUSION CONCEIVABLE

Fusion is a nuclear process in which two or more light nuclei (in stars, for example, two hydrogen nuclei) fuse to produce a single, heavy nucleus (for example a helium isotope) (for example a helium isotope). The process causes a difference in mass, which is transformed into energy. In the sun, this energy radiates away.

On Earth, if repeated, this reaction has the potential to create clean, safe energy in amounts that might solve the climate issue. So it's no wonder that governments all around the globe are investing in research to mimic the fusion process.

Nuclear fusion releases almost four million times more energy than utilizing coal, oil, or gas, and four times as much as nuclear fission technology (in which a huge atomic nucleus is divided into smaller nuclei) (in which a large

atomic nucleus is split into smaller nuclei). This implies that nuclear fusion has the potential to supply energy in proportions that could power homes, cities, and even nations.

Fuels that may be utilized for fusion are also commonly accessible and can be created from substances such as water and lithium. A few grams of fuels such as deuterium and tritium may provide a terajoule of energy - about what one person in a rich nation needs over 60 years.

Additionally, nuclear fusion does not create carbon dioxide or long-lived nuclear waste, making it a very sustainable choice. However, achieving the correct conditions in a fusion reactor has proved to be a complex and expensive procedure.

International Collaboration

Since the notion of nuclear fusion was discovered in the 1930s, tests have been continuing, and now there are roughly 20 fusion reactors around the globe, all seeking to attain

the very high temperatures required for long for fusion to happen.

In 1958 during an Atoms for a Peace meeting in Geneva, fusion research was designated a worldwide joint enterprise, resulting in the building of the largest prototype reactor in the world. This 'tokamak' - the term given to devices that produce a high magnetic field to confine reactive charged particles – was created in the South of France with support from 35 nations, including the US, EU, China, Russia, India, Japan, and the UK.

However, the International Thermonuclear Experimental Reactor (ITER) project has yet to produce power for customers, and opponents, including the European Court of Auditors, have warned the project might be prone to future cost rises and delays.

New world records

In recent months, significant advancements are making technology more accessible than ever before.

In China, in January 2022, the EAST reactor achieved the record for the longest continuous nuclear fusion with temperatures of 126 million degrees Fahrenheit - nearly five times hotter than the sun — maintained for 17 minutes.

China's EAST reactor is being used to test the technology for the ITER reactor in the south of France, with some estimates now suggesting it might begin running as soon as 2025.

Aerial image of a nuclear power plant

A month later in the UK, scientists at the Joint European Torus laboratory (JET) announced they'd generated a record-breaking 59 megajoules of sustained fusion energy – double the previous record energy output and a further boost for their partners at the ITER project (for

which it is acting as a prototype of sorts) (for which it is acting as a prototype of sorts).

Nuclear fusion technology is also growing closer to commercial usage. Entrepreneurs at Tokamak Energy, headquartered in a tiny railway town in southern England, have attracted venture funding – and £10 million from the UK government - for their space-saving and cost-effective solution. The system employs tokamaks and high-temperature superconductor magnets.

The business has indicated that it is on schedule to create commercial power from nuclear fusion by 2030.

The UK has also just initiated the STEP project (Spherical Tokamak for Electricity Production), seeking to construct a reactor that will link to the national energy grid in the 2040s.

And in the US, California-based TAE Technologies, the world's biggest private fusion

startup, has stated it would have an operational nuclear fusion power plant by 2030, after obtaining $880 million in investment.

In August 2021, the Lawrence Livermore National Laboratory revealed a huge breakthrough in nuclear fusion, employing strong lasers to create 1.3 megajoules of energy – nearly three percent of the energy contained in one kilogram of crude oil.

The initial study suggests the new findings are an eight-fold increase in those from spring 2021 and a 25-fold improvement in 2018.

Nuclear fusion technology is still some way from being ready to be scaled up for commercial usage, but innovation is proceeding at speed as we wrestle with one of the greatest issues in science. And governments are playing the long game, investing for the mid-to long-term in its exceptional potential returns.

Chapter 3

NUCLEAR FUSION ENERGY

The development of unlimited, carbon-neutral, and safe energy through nuclear fusion is expanding around the world, and scientists at the Atomic Energy Authority in the United Kingdom (AEA) have recently cleared one more key hurdle to make it a commercial reality: exhausting gas that's hotter than the Sun. The heated plasma formed during fusion power production has to cool down while it's being utilized, but at such high temperatures, there aren't any materials available to tolerate the heat. Now, that issue seems to have been fixed.

The AEA team's remedy to the heat problem is a "sacrificial wall" design which will need replacement every few years. Plasma will be pushed along a route inside its fusion generator's holding device to cool it somewhat before coming into contact with a specially built wall for the rest of the cooling process. However,

even at a lesser temperature, the heat will weaken the wall's integrity over time and need to be replaced. With the first nuclear fusion reactor expected to power on in seven years, AEA's fusion exhaust system may be one of the advances that keeps it on track.

It's claimed that imitation is the sincerest form of flattery, and current fusion energy discoveries prove that sentiment concerns don't keep within the borders of Earth. At nearly 90 million miles distant, our Sun is effectively a fusion reactor in the sky, its huge size providing enough gravity to push atoms together at its core and unleash tremendous quantities of energy. Artificially replicating the circumstances required for this form of generation is tricky, but the endeavor has been going on since the 1960s. The AEA is emblematic of one agency in a worldwide initiative.

The most advanced nuclear fusion project today is ITER, the International Nuclear Fusion Research experimental reactor in southern

France, which accommodates scientists from 35 nations devoted to producing the first-ever positive fusion energy output. Their invention is dubbed a "tokamak", and its construction is somewhat like a flattened donut (torus) enclosed by rings of strong magnetic coils. The magnetic fields produced by the coils both suspend the plasma formed by great heat and compress the plasma into a tiny region to trigger the fusion reactions. ITER is slated to switch its reactor on in 2025.

Producing fusion in a laboratory includes two basic parts: 1) creating plasma, a soup of electrons and nuclei liberated from their atomic structures owing to very high temperatures; and
2) fusing the nuclei of two distinct sorts of atoms, typically different forms of hydrogen. The heat in a tokamak is created by both the magnetic field movement and external heating equipment, and the

nuclei merge is done by compressing the plasma using the same magnetic fields into a restricted region to stimulate collisions. Essentially, the high heat

stimulates the atomic particles, quickening their travel, and their rapid motions inside the magnetically restricted space considerably increase the possibility the nuclei will smash and fuse. When nuclear fusion happens, a large quantity of energy is produced, the goal of desire for everybody interested in this area of study.

The amount of heat required to induce atoms to release their electrons and produce plasma is in the order of millions of degrees Celsius, the center of the Sun itself being 15 million degrees. Without strong gravity to help in squeezing plasma, as in the Sun's case at 27 times the gravity of Earth, reactors on our planet need to heat much beyond the Sun's temperature to guarantee the atomic particles in the plasma collide and fuse. ITER's tokamak warms to 100 million degrees Celsius.

All of this heating and magnetic control needs energy input, and here is where the present stage of fusion energy research is concentrated. The

ratio of energy required and energy generated is termed "Q", the ideal quantity strived for by scientists in the area being 10:1. When 10 times the energy is generated by nuclear fusion than consumed to make it, it will have progressed to a level fit for further development as an alternative power source, or so goes the theory. ITER's particular objective is to create 500 MW of fusion power from 50 MW of heating power.

Once energy is produced from the fusion process, it may then be harnessed to generate steam to power turbines presently utilizing other power sources such as coal and natural gas. This is another major stated benefit of fusion power; it can connect directly to current power systems, reducing any interruptions or needs for additional equipment. Combined with the plentiful supply of hydrogen and the absence of greenhouse gases or radioactive waste, there are great expectations for fusion's future as an all-in-one energy source.

Chapter 4

HOW DOES NUCLEAR FUSION WORK

To explain "how nuclear fusion works," maybe we should first ask, "how does the sun work?"

The Sun was imaged at 304 angstroms by the Atmospheric Imaging Assembly (AIA 304) of NASA's Solar Dynamics Observatory (SDO) (SDO). This is a false-color picture of the Sun seen in the extreme ultraviolet portion of the spectrum. (source)

Since the start of time, mankind has stood in awe of our sun. Ancient Egyptians regarded it as the deity Ra, who traveled through the sky in a celestial boat as one would sail down the Nile; ancient Greeks adored it as Helios, who drove a chariot from horizon to horizon driven by fire horses. Many faiths, ancient and contemporary, consider the bright, blinding disk in the sky as an emblem of celestial creatures like Aten, Utu, Tonatiuh, Sol Invictus, Ameratsu, Surya, etc. The sun provides us with heat and light, our changing seasons, and makes all life and civilization on Earth possible. You could argue that our planet revolves around the sun.

And you would be accurate because it does.

The first person in recorded history to claim that our planet revolves around the sun, physically and not only symbolically, was the Greek astronomer Aristarchus of Samos, who flourished around the 3rd century BC. Around the same time, another Greek astronomer and philosopher, Anaxagoras, proposed that the sun was not, in fact, the chariot of Helios and was instead a gigantic ball of burning metal that orbited the Earth. People did not enjoy being informed of this.

Around the same period, Eratosthenes of Cyrene, the Greek mathematician famed for measuring the circumference of the Earth with astounding accuracy, also computed the distance from the sun to the Earth as being roughly 150 million kilometers (about 94 million miles) (about 94 million miles). The sun is, in reality, 147 million kilometers distant from the Earth at the closest point in our orbit and 153 million kilometers at the furthest point.

Over the next two thousand years or so, scientists and philosophers the world over—in the Mediterranean, in the Middle East, in Asia, and Europe—learned more and more about the sun, but it wasn't until the beginning of the modern scientific era in the 19th century that we had the tools to start tackling one of the biggest questions in the world: Where does all the sun's energy come from?

Discovering Nuclear Fusion Power:

It wasn't until the 20th century, following the discovery of radioactivity, that we worked out nuclear fusion science. In 1904, Ernest Rutherford postulated that radioactive decay may be responsible for our sun's output. Soon after, Albert Einstein developed his theory of mass-energy equivalence, best expressed in his famous formula E=mc2; in 1920 Sir Arthur Eddington proposed that the sun could be producing energy, as expressed by Einstein's work, by merging hydrogen atoms to create helium and thus giving out heat and light – essentially, nuclear fusion energy production. Subrahmanyan Chandrasekhar and Hans Bethe created the theoretical notion of what Eddington had described, now known as nuclear fusion, and estimated how the nuclear fusion processes that power our sun function.

As soon as we comprehended the nuclear furnace sitting at the core of our sun, which was, in reality, a big ball of incandescent (main hydrogen) gas and not, as Anaxagoras had assumed, a blazing metal orb (nice guess, though!), we began wondering—"Hey, can we do that here on Earth, too?"

And so the hunt for nuclear fusion energy generation started.

How Does Nuclear Fusion Work? The Physics of a Nuclear Fusion Reaction

Nuclear fusion is one of the simplest, and yet most powerful, physical processes in the universe. Two tremendously excited, very hot, very energetic atoms crash with one other and convert into one atom, shedding a few residual subatomic particles and leftover energy in the process.

After the Big Bang, the whole universe was an incredibly hot, highly intense soup of very small subatomic particles—except it wasn't entirely fair to call them subatomic particles yet because atoms didn't exist at this time. Eventually, these small particles started to attract one other and connect, changing quarks into electrons, neutrons, and protons—the basic building blocks of matter. The hot, thick soup of the cosmos started to cool and curdle as it expanded, generating small lumps of hydrogen gas.

For a moment, the cosmos was nothing but hydrogen, the simplest element. But gravity eventually started to draw some of these gas clouds closer together, and as the hydrogen atoms flying about got more energy in their increasingly-dense, increasingly-hot surroundings, they began to fuse to make helium, the second-lightest element.

Many of these gas clouds became stars exactly like our sun—massive balls of hydrogen and helium plasma driven by nuclear fusion events.

And in the dense cores of these stars, hydrogen and helium fusion fuel continued to combine until it generated heavier and heavier elements. When the universe's early stars perished and burst into novas and supernovas, they flung forth clouds of all these heavier elements into space, which ultimately formed the nebulae, planets, asteroids, comets, and other interstellar bodies we know of.

How Do Neutrons Combine

It requires a significant quantity of energy—and high pressure—to cause nuclear fusion even between light atomic nuclei. Atomic nuclei, which include positively-charged protons and neutral neutrons, do not wish to get near one other under normal conditions. The Coulomb force, which defines how similar charges repel each other and opposing charges attract (as with the north and south poles of a magnet, for example), stops these two atomic nuclei from colliding with each other. If you set two atoms on a direct collision course to make their nuclei

smash into each other and stick together, you will need to accelerate them to very high speeds so that when they collide, the nuclear force, which compels protons to stick to neutrons, overcomes the repulsive Coulomb force.

Nuclear binding energy is the least amount of energy it takes to break apart an atomic nucleus. The denser the element, the more energy it takes to split its nucleus apart. When we induce nuclear fission or nuclear fusion, that nuclear binding energy may be released. This is how nuclear fission and fusion may be utilized to make power.

For heavier atoms, nuclear fusion does not emit fusion energy. But for lighter elements, such as hydrogen and helium, when two atoms combine, the resulting third atom is laden with surplus energy and an additional neutron or two in its nucleus that is making it unstable. No atom ever wants to be unstable, and thus it attempts to return to the closest point of stability by releasing all that excess. It relieves itself by

hurling off the additional high-energy neutrons, with its residual fusion energy discharged as well.

Nuclear Fusion in the Sun

In the massive fusion reactor that is the sun, the nuclear fusion process happens largely between the two light atomic nuclei hydrogen and helium, because that is the majority of its composition. As a star's life cycle continues, heavier elements develop in its hydrogen-rich core, where the mind-boggling heat and extreme pressure squeeze atoms together again and over again. Our sun is a medium-sized star at the halfway of its life cycle, having arisen from a cloud of gas some five billion years ago. Outside of its core, churning layers of heated plasma give out heat and light which travel through the abyss of space to warm all of the planets and not-quite planets (sorry, Pluto) in our solar system using nuclear fusion energy.

Eventually, around five billion years from now, the sun will deplete the once-ample supply of hydrogen and helium in its core by fusing it all into heavier elements. When that occurs, the sun will forcefully shed what remains of its outer layers and leave behind a tiny gaseous core of carbon and other heavy elements. No longer powerful enough to drive these heavy elements to combine, this surviving white dwarf will stay, dormant, in the heart of an expanding cloud of gas until it cools to become a black dwarf, no longer a source of energy for the solar system.

The sun's fusion processes occur on a scale so huge that it's tough to take it all in. In its core, the sun fuses roughly 600 million tons of hydrogen per second. It requires such a huge quantity of energy to achieve nuclear fusion that in our contemporary and mature universe, nuclear fusion will only occur spontaneously within stars like our sun, where the heat and pressure are sufficient to force light atomic nuclei together to make heavier atoms. Even hydrogen, the lightest element, needs a

tremendous energy input to fuse that just cannot naturally exist anyplace else.

And, of course, us being humans, we learned about that fusion reaction process and asked ourselves whether we could replicate it here on Earth (on a much smaller scale, of course) (on a much smaller scale, of course). After we figured out nuclear fission and created the most destructive weapons the human race has ever known, the race for nuclear fusion—as a source not of destructive power but of enough electrical fusion power to supply all of the global civilization without the need for polluting fossil fuels like coal or oil—began.

How Do Nuclear Reactors Work

There are two broad categories of nuclear reactors: nuclear fission reactors, which split heavy atoms apart into less-heavy atoms to produce byproducts such as neutron radiation, radioactive waste, and most importantly, an excess amount of energy released that can be

converted to electricity to power our homes and industries; and nuclear fusion reactors, which combine light atoms into less-light atoms in a fusion reaction to produce byproducts such as neutron radiation and (in theory) excess fusion energy production. Nuclear fusion as a source of energy production—fusion power—is the holy grail of fusion physics.

Fusion vs. Fission

A nuclear fission reactor utilizes uranium as fuel. When a uranium atom gets excited and destabilized by exposure to neutron radiation, it breaks apart into smaller atoms such as barium and krypton, and produces additional neutron radiation, which in turn excites and breaks apart other uranium atoms, generating a chain reaction. The energy released causes the water in the reactor to boil, changing into steam and powering a turbine, which subsequently creates electricity. Some of the lighter elements created in these chain events are very radioactive and require tens of thousands of years or more to

decay, making disposal hard. It's also possible for nuclear fission reactors to release radioactive material into the environment if the chain reaction gets out of control, as what happened in nuclear power plants at Chornobyl and Three Mile Island; this dangerous reaction results in an escalating release of heat and radiation, an occurrence that is only possible with fission vs fusion which cannot experience runaway reactivity. Modern fission reactors in nuclear power facilities are equipped with very redundant safety and cutoff mechanisms to avoid these types of catastrophic situations.

Not all nuclear fission reactors are nuclear power facilities intended to create electricity. Non-power-generating research reactors are employed for their neutron output for purposes like radiation survival testing, neutron radiography, and medicinal isotope synthesis.

A fusion reactor is a completely different beast from a fission reactor. For starters, a fusion reactor operates with considerably lighter

materials as a fusion fuel. In the sun, we mostly witness hydrogen, the lightest atom, fused to generate helium, the second-lightest element, with a corresponding release of fusion energy. Here on Earth, fusion reactors combine light elements such as deuterium and tritium, two hydrogen isotopes, as fusion fuel. Historically, fusion research has worked mostly with deuterium and tritium as potential nuclear fusion fuel instead of imitating the hydrogen-hydrogen and helium-helium fusion processes like our sun. This is because while the sun's method works fine due to its gargantuan mass and size, at our much more modest scale using nuclear fusion devices, we can more easily induce a fusion reaction with a deuterium atom colliding with another deuterium atom (or tritium atoms) than with a hydrogen or helium fusion reaction.

Nuclear Fusion Reactions, Nuclear Byproducts, and Radioactive Half-Life:

One of the biggest possible advantages of nuclear fusion power plants over fission, and what makes fusion such an appealing source of electrical power compared to not just fission but also essentially every other energy source, is the waste material wasted fusion fuel it leaves behind. Nuclear fission reactors leave behind extremely heavy elements from the breaking of uranium atoms which stay highly radioactive for up to tens or hundreds of thousands of years.

Every unstable and radioactive isotope has a "half-life," or the length of time it takes for half of any given sample of the substance to decay into a stable isotope that is no longer radioactive. For example, uranium-235, the specific isotope of uranium used as nuclear fuel, has a half-life of almost seven hundred million years, whereas molybdenum-99, an isotope used to make contrast agents for medical imaging, has a half-life of around two and a half days.

A buffet of radioactive waste byproducts is formed by uranium and plutonium fission, some

of which have half-lives of days or hours and others of which have half-lives over two hundred thousand years. How to store and dispose of long-lived nuclear waste is a serious challenge related to fission power, but a nonissue in fusion power. Deuterium-deuterium and deuterium-tritium reactions create helium-3 and helium-4, two stable isotopes of helium.

How a Nuclear Fusion Reactor Would Work to Produce Fusion Power

There are two basic kinds of fusion reactor designs, both of which are being explored in modern fusion research: magnetic confinement reactors and inertial confinement reactors.

Magnetic Confinement Reactors

Fusion reactions begin with heated plasma, the fourth basic state of matter. Plasma is a hot, electrically conductive plasma of ions and unbound charged particles that provide the ideal

crucible for nuclear fusion, and all of our technology utilized to induce fusion requires wrangling and regulating this state of matter in a high-energy, high-intensity environment. When two light atomic nuclei hit one other at fast speeds, they may more readily overcome the Coulomb barrier and fuse, releasing the ions' nuclear binding energy. This is what occurs at the core of our sun. To reproduce that energy-creating fusion process in a fusion reactor here on Earth and harness fusion power for our use, we need equipment that regulates the flow of superheated plasma.

How Do Magnetic Confinement Reactors Work

A magnetic confinement fusion system depends on employing intense magnetic fields to contain and regulate the flow of superheated plasma. As particles inside the plasma are steered by a strong magnetic field, they smash with one other

and fuse forming new elements. The notion of magnetic energy confinement for a fusion reactor was initially proposed in the 1940s, and early fusion research left scientists confident that magnetic confinement would be the most viable technique to create fusion energy.

The most well-explored and well-known form of magnetic confinement system is the tokamak reactor, initially designed by Soviet scientists Igor Tamm and Andrei Sakharov in the 1950s based on Z-pinch machines. A tokamak is a doughnut-shaped fusion reactor that creates a helix-shaped magnetic field using massive electromagnets situated in the inner ring. A comparable fusion reactor concept, termed a stellarator, employs external magnets to produce a containment field to the superheated plasma inside the fusion reaction chamber. The major distinction between a tokamak and a stellarator's fusion reactor architecture is that a tokamak depends on the Lorentz force to twist the magnetic field into a helix, while the stellarator twists the torus itself.

In the 1970s, and with a surplus of funds coming into research institutes from governments to build fusion power plants to address energy demands during the oil crisis, experimental tokamak and stellarator (but largely tokamak) fusion reactors started to crop up all over the globe. It didn't take long to learn that magnetic confinement fusion, although capable of creating clean fusion power, was far more difficult to carry off than planned. To kick-start a reaction with a fusion power output of more fusion energy than it takes to sustain it and then keep it running (which is the important thing), you need very powerful magnets to keep the plasma flowing smoothly through the tokamak fusion reactor's ring, and you need to sustain very high temperatures—temperatures over one hundred million degrees—to make up for having less pressure in the reactor than there is in the intense pressure of the sun's core.

While the United States part of the fusion experiment financing dried up in the mid-80s

after then-president Ronald Reagan proclaimed the energy crisis ended, work on tokamak development for fusion research continued. Design work started on ITER, or the International Thermonuclear Experimental Reactor, in 1988. This was a combined endeavor of fusion science researchers from the United States, Soviet Union, European Union, and Japan since fusion energy experts soon realized that no one country could create a strong enough tokamak fusion reactor on their own.

Similar to ITER is the Joint European Torus, or JET, situated at Culham Centre for Fusion Energy in the United Kingdom. The Joint European Torus is the world's biggest operating magnetically confined plasma physics experiment and one of its principal present functions is to test and enhance aspects of ITER's design. JET is one of the few facilities in the world that produces more neutrons than us!

Currently, while advances in plasma science and materials science are still needed to make fusion

reactors that can output more fusion energy than it takes in, tokamak reactors are still regarded as the most promising path in fusion research to one day create power plants for clean fusion energy production.

Inertial confinement fusion depends on blasting high-energy laser beams at a fuel pellet target holding deuterium and tritium fuel for the reaction. The collision of the high-energy beam produces shockwaves to propagate through the fuel pellet target, heating and compressing it to initiate fusion reactions.

This method of inducing nuclear fusion reactions this way and harnessing inertial fusion energy was first suggested in the 1950s, and in the 1970s, high-energy ICF (inertial confinement fusion) research suggested that it could be a more promising path to fusion energy than tokamak and stellarator fusion reactors.

However, during the following two decades, fusion power experts steadily uncovered more

and more barriers that needed to be overcome to accomplish ignition inside such a fusion reactor, and estimates for how much energy the laser beams required to induce fusion quadrupled on an annual basis.

The National Ignition Facility at the Lawrence Livermore National Laboratory in Livermore,

California is the biggest and most energetic ICF system in the world. Completed in 2009, as of 2015, this system has only been able to attain one-third of the parameters required for ignition. The NIF is now utilized mostly for materials science and armament research rather than fusion power development.

Other Places Where Nuclear Fusion Happens Nuclear fusion processes only naturally occur in stars, but here on Earth, nuclear fusion isn't simply occurring at ITER and other fusion energy research labs. There are additional nuclear fusion research centers studying fusion projects such as colliding beam fusion, which involves speeding a beam of high-energy

particles into a stationary target or another beam to generate a nuclear fusion reaction, similar to inertial confinement fusion. While this artificial fusion experiment doesn't have as much promise for fusion power production at the time, it has other benefits in science and industry that are no less essential than fusion research.*

Nuclear fusion also happens within thermonuclear or fusion weapons, often known as hydrogen bombs, which every sensible person on Earth hopes we never, ever, ever have to use.

As it turns out, one of the most immediately valuable outputs of fusion reactions—particularly deuterium-deuterium and deuterium-tritium reactions—isn't energy, but rather neutron radiation. Neutron radiation is a consequence of all nuclear processes, including fission and fusion, and since the 1950s, industrial and scientific uses such as neutron radiography and medical isotope synthesis have relied on fission reactors for their high-energy neutron output. But recent breakthroughs in

colliding beam fusion, or accelerator fusion, are making fusion a more practical approach to manufacturing neutrons than fission.

On the biggest scale of the colliding beam, fusion are giant particle accelerators such as the Spallation

Neutron Source at Oak Ridge National Laboratory, which creates tremendous neutron yields and is mainly utilized for neutron scattering studies. Scientists utilize neutron scattering to better understand the molecular makeup of materials such as metals, polymers, biological samples, and superconductors.

On the smallest scale of a colliding beam, fusion is sealed-tube neutron sources, which are very small accelerators—small enough to fit on a table or workbench, and often small enough to be used for fieldwork—that work by shooting particle beams of deuterium or tritium ions at a deuterium or tritium target to make the target undergo fusion. The smaller the neutron source, the lower its yield, and therefore compact sealed-tube sources tend to be employed largely

for work that only requires a low neutron yield from a portable source, such as oil well logging, coal analysis, and most applications of neutron activation analysis. These sealed-tube sources are commonly employed in the petroleum sector.

between enormous spallation sources and small sealed-tube neutron sources are SHINE's high-flux neutron producers. Our high-flux neutron generators function under the same fundamental principles as sealed-tube sources, only enormously scaled up from tabletop-sized neutron emitters so that they may be employed in the same high-yield industrial and research niches as conventional fission reactors. Our devices depend on inertial electrostatic fusion, not magnetic confinement fusion—meaning that the plasma is confined by a strong electric field, not a magnetic field.

Our sister company Phoenix's imaging center in Fitchburg, Wisconsin uses a high-yield neutron source to perform neutron radiography, which is crucial for aerospace manufacturers; meanwhile,

at our campus in Janesville, Wisconsin, our medical isotope production facility will use our accelerator-based nuclear technology to produce potentially lifesaving nuclear medicine for diagnostic and therapeutic procedures. Utilizing our fusion technology, we anticipate our Mo-99 manufacturing plant will be capable of supplying almost a third of the world's demand for molybdenum-99 in the next years, which permits tens of thousands of diagnostic operations every day!

Will Nuclear Fusion Research Give Us a Future of Fusion Energy

Coming back full circle to humanity's ambition to master the power of the sun, high-yield fusion neutron sources, albeit ill-suited to produce the scientific holy grail of a fusion power plant, may be employed to assist us to reach that objective. To make controlled fusion power as a source of energy a reality, we need stronger materials to

use in a nuclear fusion system and reactors, such as superconducting magnets and shielding material that can withstand the intense operating conditions, and through techniques such as neutron scattering and radiation hardening, we can further fusion research to design and develop the reactor for the fusion power plant of tomorrow.

SHINE's Vision for Fusion as a Source of Energy
SHINE's four-phase vision lays out a practical, stepwise approach to improving fusion technology and overcoming the technological barriers to fusion power by exploring near-term and long-term applications of nuclear fusion, including advanced industrial inspection, medical isotope production, and nuclear waste recycling. We think that our method of expanding nuclear fusion is the greatest hope we have at making the ultimate in atomic energy a reality for global civilization—and beyond.